**For Eddie, Olivia, Lucy, and Eamon, who make
all of my days into holidays.
—HB**

**For my lovely family.
—CA**

Text © 2025 by Hannah Barnaby
Illustrations © 2025 by Cédric Abt
Cover and internal design © 2025 by Sourcebooks
Images © Churkin Llya/Shutterstock, Hannah Barnaby

Sourcebooks and the colophon are registered trademarks of Sourcebooks.

All rights reserved.

No part of this book may be used or reproduced in any manner for the purpose of training artificial intelligence technologies or systems.

The characters and events portrayed in this book are fictitious or are used fictitiously. Any similarity to real persons, living or dead, is purely coincidental and not intended by the author.

All brand names and product names used in this book are trademarks, registered trademarks, or trade names of their respective holders. Sourcebooks is not associated with any product or vendor in this book.

The full color artwork was created using Photoshop on a Huion Kamvas Pro tablet.

Published by Sourcebooks eXplore, an imprint of Sourcebooks Kids
P.O. Box 4410, Naperville, Illinois 60567-4410
(630) 961-3900
sourcebookskids.com

Cataloging-in-Publication Data is on file with the Library of Congress.

Source of Production: China
Date of Production: February 2025
Run Number: 5043477

Printed and bound in China.
TL 10 9 8 7 6 5 4 3 2 1

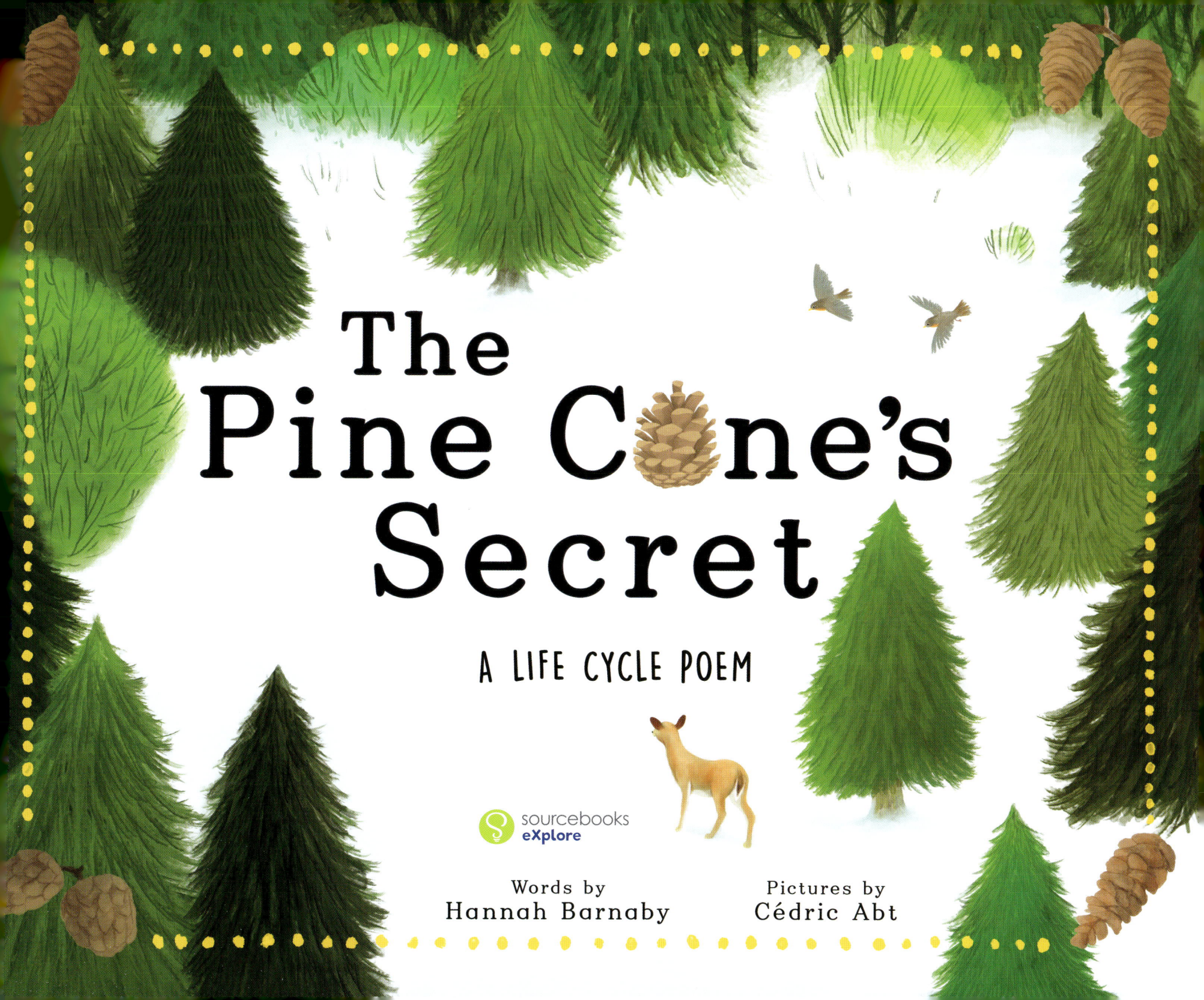

A pine tree is a cone.
Clinging and swinging and hanging up high,
Sealed like a secret and ready to fly.

The wind starts to blow,
Then look out below!
A pine tree begins with a cone.

A pine tree is a sapling.
Stretching and reaching for sunlight and rain,
Watching the seasons pass once, twice, again.

Roots digging down,
Her branches a crown,
A pine tree turns into a sapling.

A pine tree is a tower.
Covered in needles, bristly and lean.
Steady, unchanging, a coat evergreen.

Patiently growing,
Trusting and knowing
Someday, she'll be a tall tower.

A pine tree is a haven.
Each seed tucked in tight
From the cold, wet, and wind,
While the pine cones are
Waiting to open again.

Where the scales overlap,
Little seeds take a nap.
A pine tree provides a safe haven.

A pine tree is a magic trick.
Come summer or dry air or moisture or fire,
She knows what conditions her seedlings require

To take hold and thrive,
Keep the forest alive.
A pine tree performs her own magic.

A pine tree is a feast.
As the pine cones unfurl
When the summer is done
To set their seeds free,
They don't drop every one.

And the nuts they hold back
Make a fabulous snack.
If you're lucky, a pine tree's a feast!

A pine tree is a home.
Green when the other trees' leaves disappear,
A pine says to critters, "Come, warm up in here."

Plenty of space
It's a warm, cozy place.
A pine tree is a welcoming home.

A pine tree is a blanket.
When the temperature drops and the animals rest,
Its needles and cones make a soft quilted nest.

Under thick insulation,
In deep hibernation.
A pine tree prepares a warm blanket.

A pine tree is a gift.
Make wreaths from her branches,
A garland or two.
Smell the spicy sharp scent that
The pine shares with you.

Roll a pine cone in seeds.
See how many it feeds!
A pine tree's a generous gift.

Ornaments, lights,
She loves anything bright.
A pine tree is your celebration.

A pine tree is a family.
Each tree from before
made the trees we have now,

From their countless pine cones sending seeds to the ground.

Just like you, and like me,
And the whole world we see...

A pine tree is part of a family.

PARTS OF A PINE TREE

LONG TIME, NO TREE!

Once a pine seed finds a safe place in the ground, it takes a long time to grow! Different types of pines take anywhere from five to thirty years to be old enough to grow cones. It takes two to seven years for cones to form, and two or three more years for the seeds inside the cones to get ready to make new trees. It's a long process!

A Scotch Pine like the one in the story grows like this:

Throughout the Great Lakes region and the Northeast, a Scotch Pine can grow 2½ feet per year. That's 12 to 15 times faster than you!

Just like kids, some Scotch Pines grow faster than others—it all depends on where the seeds came from, how much space and sunlight they get, and whether they have to fight off pests.

2 years old:
8 to 14 inches

5 years old:
10 feet

21 years old:
45 feet

75 years old:
125 feet

THE NEED FOR SEEDS

Once pine cones are ready to release their seeds, they might need some help. Seeds can't get very far on their own—they don't have legs or cars. So how do seeds get to new places to grow into new trees?

WEATHER

Wind can carry seeds, some of which have special features that help, like fluff or wing-like parts that catch the breeze.

Rain can knock seeds to the ground and form tiny rivers that carry the seeds to new locations.

ANIMALS

Forest creatures eat plants or seeds and travel somewhere else in the woods. When they poop, the seeds come out!

Some seeds grow inside pods that have little hooks, which catch onto animals' fur or feathers and get carried away.

PEOPLE

If you go for a walk or a hike in the forest, you might find some of those sneaky little seed pods attached to your socks or your jacket.

Some people harvest pine nuts from trees or seeds from plants, and plant them somewhere new.

BRANCHING OUT

Trees from the Pinus family can be found throughout the Northern Hemisphere, but they're not all the same. More than 120 types of pine trees grow north of the equator! Here's a list of some varieties and where they grow, in order from smallest to tallest.

 The Rocky Mountain Bristlecone Pine (8–30 ft.) is native to the Rocky Mountains (Arizona, Colorado, and New Mexico). It may be one of the shortest pine trees, but it's also one of the oldest: Most of the Great Basin Bristlecones in Nevada are over three thousand years old.

 The Pinyon Pine (10–20 ft.) is a short, scrubby tree with small cones that grow loads of seeds. Various Native American tribes in the Southwestern United States collect and eat these pine nuts.

 The Eastern White Pine (50–100 ft.) is also called the Great Tree of Peace by the Haudenosaunee nations. It's one of the most common choices for a Christmas tree in North America.

 The Pond Pine (30–70 ft.) is one of the species that holds its cones closed for many years, opening them to drop seeds only after a wildfire has scorched them. It's mostly found in the Eastern United States.

 The Mexican Weeping Pine (60–80 ft.) grows drooping tufts of long needles and originates in Mexico.

 The Sugar Pine wins the prize for tallest tree! Found in California, Nevada, Oregon, and Mexico, Sugar Pines can grow up to two hundred feet tall, and they grow the longest cones, nearly two feet in length.

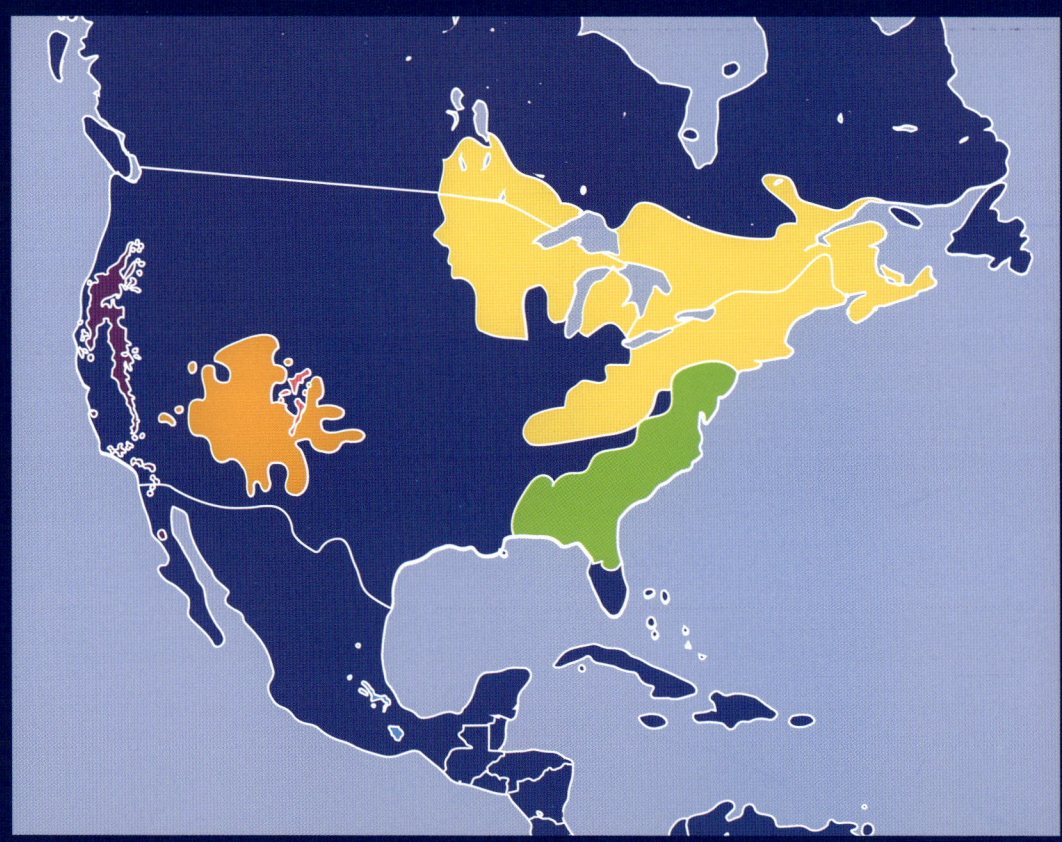

 The Italian Stone Pine (30–60 ft.) grows in an umbrella shape! It's native to southern Europe, Lebanon, and Turkey.

 The Austrian Pine, aka the **European Black Pine** (40–100 ft.), is native to southern Europe, northern Africa, Cyprus, and Turkey.

 The Japanese White Pine (20–50 ft.) is a favorite of bonsai artists. It grows in a flat, spreading pattern with purplish-brown bark. It originates from Japan and South Korea.

 The Aleppo Pine, aka the **Jerusalem Pine** (30–60 ft.), grows in hot climates and has orange-red bark. It grows best in climates like the Mediterranean.

 The Chir Pine is native to the Himalayan regions of Asia (Afghanistan, Bhutan, China, India, Pakistan, and Nepal). Bunches of long needles give it a fluffy appearance, and it can grow up to 180 feet.

TREE TRIVIA

 Pine trees can live for hundreds or even thousands of years. In eastern California, one Great Basin Bristlecone Pine named Methuselah is almost five thousand years old, according to tree-ring data. That makes the tree the oldest living thing on land!

 Red squirrels are notorious for collecting green pine cones and saving them for the winter. They gather huge piles of food called middens that can keep them and their families fed for several seasons. One group of red squirrels hid fifty pounds of pine cones under the hood of a car in Michigan in 2018.

 The cones of the Coulter Pine are the heaviest in the world. They can weigh over ten pounds. Look out below!

 The Jeffrey Pine is known as "gentle Jeffrey" because the prickles on the end of its pine cone scales are turned inward, so they won't poke you. (But be careful: The "Prickly Ponderosa" looks like the Jeffrey, but its scales stick out. How rude!)

 In 2011, two men discovered the world's tallest pine tree in southern Oregon. It's a Ponderosa Pine ("Prickly Ponderosa") that lives in the Rogue River-Siskiyou National Forest, and it's over 268 feet tall. That's just a little bit shorter than the Statue of Liberty and her pedestal!

 Every year, a huge pine tree is transported to Rockefeller Center in New York City to become the official Christmas tree of the city. The largest Rockefeller tree to date was delivered in 1999 from Killingworth, Connecticut, and it stood one hundred feet tall.

SOME SAPPY HUMOR

What do you call a pig in an evergreen forest?
A porkypine!

Why did the pine tree get in trouble?
Because it was being knotty.

Why do pine trees hate hard questions?
Because they get stumped!

PINE CONE CRAFT

MATERIALS NEEDED:

- Styrofoam eggs (available at craft stores)
- Brown paper grocery bag or construction paper
- Sharp scissors
- Low-temp glue gun
- Marker or pencil

STEP 1:

Ask a grown-up to help cut your grocery bag down one side and around the bottom edges. Spread the bag flat on a table or the floor. Cut the bag into several strips about 1 inch wide and 8 to 12 inches long.

STEP 2:

With a grown-up's help, cut a series of semicircles along one long edge of each strip of paper. You can draw the pattern with a marker or pencil first. This creates a scalloped edge on one long side of each strip. Make sure not to cut too deeply! You want to leave enough paper at the base of the strip, so the Styrofoam egg doesn't show through when you layer your pine cone's scales.

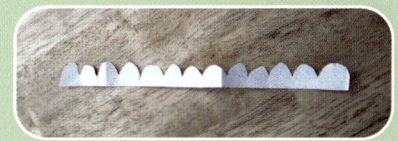

STEP 3:

With a grown-up's help, make an indentation in the top of your egg using the tips of your scissors. You can also use your pencil to make the indentation.

STEP 4:

Roll the end of one of your paper strips into a little spiral. Ask a grown-up to put a dot of hot glue in the indentation you made at the top of your egg. Secure the rolled end of the paper in place.

STEP 5:

Ask a grown-up to put a small line of hot glue around one side of the egg and attach the straight edge of your paper strip. (The scalloped edge should point up!) Wind the strip around the egg along the glue line to attach it. You may need to crimp or pinch the paper in places, so it lays flat.

STEP 6:

Keep adding lines of glue and attaching paper strips in an overlapping spiral pattern, winding toward the bottom of the egg until all the Styrofoam is covered.

STEP 7:

To finish off the bottom of your pine cone, cut a small circle of paper and glue it to the base of the egg.

STEP 8:

If you want to hang your pine cone on your Christmas tree or elsewhere in your house, ask a grown-up to help you pin or glue a loop of ribbon, yarn, or twine to the top of the egg. If you have some small wooden or glass beads, tuck them into the paper scales of your pine cone to look like seeds that are waiting to grow!

HANNAH BARNABY lives with her family near the Blue Ridge Mountains in Virginia, where there are many varieties of pine trees and cones. Her favorite is the Loblolly Pine because it's the most fun to say. *The Pine Cone's Secret* was inspired by a Christmas ornament in the shape of a pine cone that she hangs on her tree every year.

CÉDRIC ABT is a French illustrator whose work is a mix of traditional and digital techniques. For Cédric, the process is a sort of laboratory where experimentation and accidents give life to his drawings and create dreamlike and colorful universes. His creative journey is based on themes that generate inspiration: human nature, environmental nature, childhood, and more. He lives in Brittany in northern France, closest to nature and the sea.